The Open University

Technology Foundation Course Unit 31

THE ECONOMICS OF TRAFFIC CONGESTION

Prepared by Ray Thomas
for the Technology Foundation Course Team

THE OPEN UNIVERSITY PRESS

Technology Foundation Course Team

G. S. Holister (*Chairman and General Editor*)
K. Attenborough (*Engineering Mechanics*)
R. J. Beishon (*Systems*)
G. Bellis (*Scientific Officer*)
T. D. Bilham (*Course Assistant*)
D. A. Blackburn (*Materials Science*)
J. K. Cannell (*Engineering Mechanics*)
A. Clow (*BBC*)
G. P. Copp (*Assistant Editor*)
C. L. Crickmay (*Design*)
N. G. Cross (*Design*)
E. Goldwyn (*BBC*)
J. G. Gregory (*Editor*)
J. G. Hargrave (*Electronics*)
R. D. Harrison (*Educational Technology*)
R. Hermann (*Scientific Officer*)
M. J. L. Hussey (*Engineering Mechanics*)
A. B. Jolly (*BBC*)
J. C. Jones (*Design*)
L. M. Jones (*Systems*)
J. McCloy (*BBC*)
D. Nelson (*BBC*)
C. W. A. Newey (*Materials Science*)
S. Nicholson (*Design*)
G. Peters (*Systems*)
A. Porteous (*Engineering Mechanics*)
C. Robinson (*BBC*)
R. Roy (*Design*)
J. J. Sparkes (*Electronics*)
L. A. Suss (*Academic Administration*)
R. Thomas (*Economics*)
M. Weatherley (*BBC*)
G. H. Weaver (*Materials Science*)
P. I. Zorkoczy (*Electronics*)

The Open University Press

Walton Hall Bletchley Buckinghamshire

First published 1972

Printed in Great Britain by
Martin Cadbury Printing Group

SBN 335 02534 X

Open University courses provide a method of study for independent learners through an integrated teaching system, including text material, radio and television programmes and short residential courses. This text forms part of a series that makes up the correspondence element of the Technology Foundation Course.

The Open University's courses represent a new system of university-level education. Much of the teaching material is still in a developmental stage. Courses and course materials are, therefore, kept continually under revision. It is intended to issue regular up-dating notes as and when the need arises, and new editions will be brought out when necessary.

For general availability of supporting material referred to in this book, please write to the Director of Marketing, The Open University, Walton Hall, Bletchley, Buckinghamshire.

Further information on Open University courses may be obtained from the Admissions Office, The Open University, P.O. Box 48, Bletchley, Buckinghamshire.

Contents

Aims and Objectives

The aim of this unit is to provide an account of the essential ideas of the economics of traffic congestion. The text of the unit is short because the essential ideas are few in number. But within the unit (and parts of the file with it) is included ancillary material in the form of graphs, statistics and short quotations which support, or in some cases controvert, the argument given in the unit.

After working through the unit you should be able to:

*1 Define private, social and public costs.

*2 Explain the existence of traffic congestion in terms of the distinction between private and public costs.

3 Explain why the growth of road vehicle ownership does *not* often lead to traffic coming to a standstill even in the most congested areas.

4 Recall the form of the relationship between the speed and flow of vehicles on a typical urban road network.

*5 Define the marginal public cost of an extra vehicle on the road system and (given data on vehicle running costs and 'desired' speed) calculate the marginal public cost at different traffic speeds.

6 Discuss the principles involved in taxation of the motorist and the financing of road construction.

*7 Evaluate alternative solutions to the problem of traffic congestion.

What you have to do

Objectives 1 to 5 inclusive can be met by study of the main text of this correspondence unit alone. There are exercises at the end of Sections 2, 3, 5 and 6, which will help you to test whether you have achieved these objectives.

Objectives 6 and 7 are different in character. There are few generally accepted principles according to which taxes should be levied and few principles governing the precise way in which road construction and maintenance should be financed. But this unit should add to your ability to play a part in discussion of such topics.

Neither are there universal solutions to the problem of traffic congestion, or *correct* answers to the questions you might be asked on this topic. You will be expected to take a view on solutions to the problem of traffic congestion and to support this view with argument and evidence drawn from the relevant parts of text of the correspondence unit, and from the ancillary material, the file, and other sources such as the transport case study in the Book of the Course.** You will also be expected to deal with relevant evidence or argument given in this material which controverts your view. But you will not gain credit in answering any question on this topic for the particular views you put forward. You will be assessed solely on the skill you show in presenting evidence and argument to support your view.

* Questions based on these objectives may be included in the terminal examination.

** The Open University (1972) *The Man-made World* The Open University Press.

Section 1

1 Introduction

Many different kinds of answers can be given to the question: 'Why does traffic congestion occur?' One answer can be given in technological terms. Traffic congestion can be said to exist because the rate of progress in the design and production of road vehicles has been much more rapid than the rate of progress of the road system to accommodate these vehicles. An answer in these terms is in many ways very appropriate for Britain since this country enjoys the distinction of having more vehicles per mile of road than any other country in the world.

This general kind of answer, however, does not take us very far. Britain may have the highest density of vehicles per mile of road space but it is also true that not all roads in Britain are congested. Most congestion occurs on a small part of the total road system—on the major streets in urban areas. The question 'Why does traffic congestion occur?' can then be answered in more sophisticated terms by pointing out that the structure of Britain's urban areas was created at a time when public transport was the dominant mode of transport, and that this structure is unsuited to a situation where more than half of households possess a motor-car.

Country	vehicles per road mile 1969
USA	28·6
Canada	15·8
Australia	8·4
New Zealand	17·5
India	1·7
Japan	24·7
EUROPE	
Belgium	39·0
Denmark	33·5
Finland	16·8
France	28·0
Germany (West)	55·5
Great Britain	62·6
Irish Republic	12·0
Italy	56·1
Netherlands	57·3
Norway	19·2
Sweden	21·8
Switzerland	38·8

Figure 1 *Cars, goods vehicles, buses and coaches (only) per road mile.*

A very specific answer to the question could then be given in terms of *queueing theory* (which is usually regarded as a branch of statistics or operational research). The number of vehicles using some parts of the road system is greater than the capacity of these parts of the system. As a result queues form, vehicles delay each other, and we get the phenomenon called traffic congestion. Queueing theory helps to explain the relationship between the number of road vehicles, the capacity of the road system and the severity of congestion.

Distribution of motor vehicles on roads in Great Britain 1966–7

Percentage of road-km	Motor vehicles per day exceeded on these roads	Percentage of motor vehicle-km
Busiest 1%	17 000	16
5	7 000	46
10	3 500	63
25	1 200	84
50	360	95
75	110	99·1
90	40	99·9

Source: Janice A. Timbers (1968), *Traffic survey at 1 300 sites*, Road Research Laboratory Report LR 206, London, HMSO.

Economists might place such a queueing theory analysis in a general context by invoking *the law of diminishing returns*. The law of diminishing returns states that successive additions of one factor of production to fixed quantities of other factors will beyond a certain point result in successively smaller increments of output.* To explain the existence of traffic congestion the law of diminishing returns might be rewritten as: successive additions of vehicles to a given road system will lead to successive reductions in speed. But this law doesn't seem to be particularly useful in the context of traffic congestion because the nature of the appropriate measure of output is not clear. Speed by itself is not an adequate measure of output. One alternative measure of output is, for example, the flow of traffic (i.e. vehicles per hour). But if the flow of traffic is taken as a measure of output as well as being an input then the problem of traffic congestion appears in a rather different light. Maximum flow is achieved at low speeds which are regarded as indicating rather congested conditions.

None of these explanations of traffic congestion gives much hope of finding practical solutions. They all point in the same direction: that if the growth of vehicle ownership is to be accommodated we need major improvements in the road system. But it is a very costly and slow process to adapt our urban areas to high car-usage levels, and if the experience of Los Angeles and Detroit can be taken as a guide no one can be confident that extensive new road building will in fact eliminate congestion.

> . . . even with all the advantages that their circumstances provide for the success of a motorway policy, many Americans are coming to doubt whether it provides a final solution. Each new motorway, built to cope with existing traffic, seems to call into existence new traffic sufficient to create a new congestion. The leading case is that of San Francisco and its satellite cities, where the citizens recently voted to tax themselves for a new system of inter-urban railways, because the motorways were becoming too congested.
>
> HMSO (1963) *Traffic in Towns—a study of the long-term problems of traffic in urban areas* (The Buchanan Report), London, HMSO, para. 23 of Report of Steering Group.

* For a discussion on this point see: The Open University (1971) *Understanding Society*, Unit 11, Production and Supply, The Open University Press.

These considerations show that it is of limited usefulness to view traffic congestion in a mechanistic way as the interaction of vehicles and the road system. To explain traffic congestion in terms which will have clear implication for policy it is more fruitful to look at the decisions made by the individual vehicle user, and to explain why people use vehicles in spite of the frustrating conditions produced by heavy and slow-moving traffic. This unit is therefore mostly about a recently developed branch of economics which provides an analysis of the existence of congestion in terms of the rational behaviour by the individual vehicle user.

An explanation in terms of the economics of congestion leads on to the second major question of the unit: 'What sort of solutions should we adopt to solve the problem of traffic congestion?' It has already been suggested that it might be possible to reduce congestion by building new roads, but if the existing urban road system is taken as fixed (as we are obliged to do in the short run) what kind of measures can be effective in reducing the severity of congestion, or at least in preventing congestion becoming more extensive?

Attempting to answer these questions involves a consideration of a variety of factors—such as the alternatives available to those who use vehicles in congested conditions, the practicality of different methods of restricting vehicle use, and of different methods of encouraging people to use means of transport other than the private road-vehicle. The economics of traffic congestion does not provide definitive answers to these questions, but it does provide a theoretical framework for analysing the advantages, disadvantages and likely consequences of different measures.

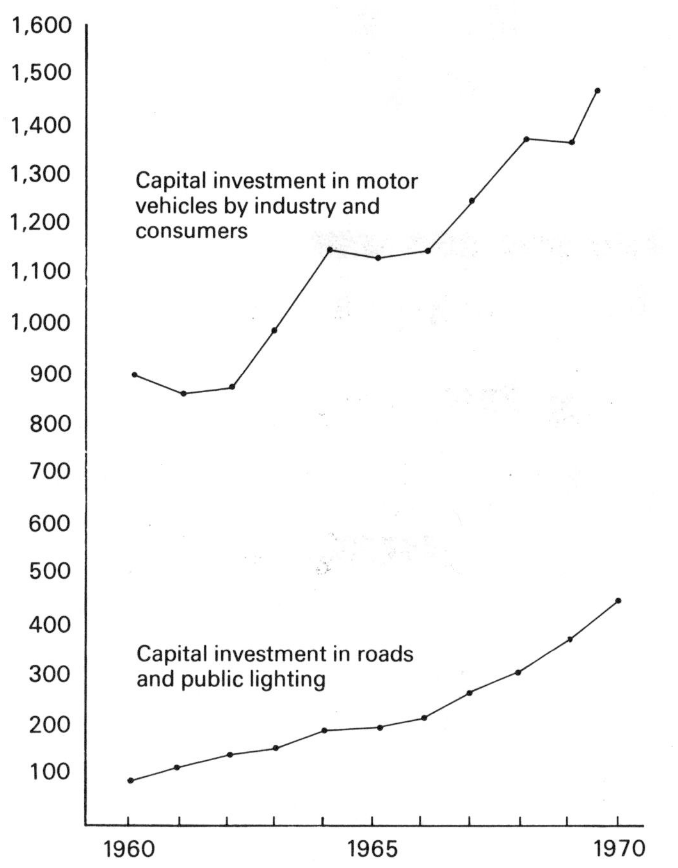

Figure 2 Capital investment in road vehicles and roads in the United Kingdom 1960–70.

Source: HMSO (1971), *National Income and Expenditure 1971*, London, HMSO.

2 Private, social and public costs

Let me start by defining terms as they will be used in this unit. *Private* costs are those borne by an individual as he carries out a particular activity. *Social* costs are the total costs of that activity to all members of the community, including those borne by the individual carrying out the activity. *Public* costs are the difference between private and social costs, or in other words they are the costs which an individual's activity imposes upon other members of the community. The relationship between these three costs can be expressed algebraically as:

$$s = p_r + p_u$$

where s = social cost, p_r = private costs and p_u = public costs.*

Private Cost

Public Cost

Private Benefit

Public Cost

The classic example of public cost is the factory chimney belching out black smoke. The private costs involved are part of the fuel bill of the factory owner. But the public costs are borne by the people who live near the factory who have, for example, to spend more money in soap and other cleaning agents than they would if the factory chimney did not deposit grit on their homes.

* Words which are part of everyday speech are commonly given precise and slightly new meanings in the language of the social sciences. *Private costs* is used here more or less in the same way as in ordinary usage except that the term is not restricted to financial costs. The term *social costs* is part of the language of economics (see The Open University (1971) *Understanding Society*, Unit 3, The economic basis of society, p. 77, and Unit 12, Markets and prices, p. 84, The Open University Press). The term *public costs* as used here appears to have become established in the field of the economics of congestion.

Almost any kind of activity which has a deleterious effect on the environment can be regarded as an example of public costs. The brewer who sells his beer in cans incurs extra packaging costs. But the costs of dealing with the empty cans do not fall on the brewer. They are met by the various public authorities concerned with keeping public places tidy and with refuse disposal. Chemical firms pay for foaming agents in detergents sold for domestic use. But the problems caused by foam in water works are dealt with by the sewerage departments of local authorities and the costs are ultimately met by the ratepayer.

In all of these examples at least some part of the costs involved are financial. But in many cases public costs take a subjective form where no money expenditures are incurred. The flat dweller who loudly plays his transistor radio or tape recorder in the early hours of the morning is a nuisance to his neighbours. Although no financial transactions are likely to be involved the nuisance is real enough and can usefully be regarded as a public cost.

Consider an example which is rather closer to the traffic congestion situation. Suppose that an individual enters a lift containing six people and that the lift thereby becomes uncomfortably overcrowded. In entering the lift the individual can be said to suffer a certain subjective cost because of the crowding. Let it be assumed that the subjective cost to the individual equals x.

Private Cost

Now the individual is not the only person to suffer from the crowding. There are six other people in the lift. If it is assumed that the subjective cost of crowding to each of these six people is also equal to x, then the public cost of the seventh individual's action in entering the lift is $6x$. As in all the other examples we have given the social cost of the activity is very different from the private cost.

The obvious method of dealing with social costs is by regulation. The Clean Air Acts normally deter the factory from belching out black smoke. The problem of overcrowded lifts is not very severe because there are usually regulations of the form '*Not more than six persons may use this lift*'. (Regulations of this kind may be justified in terms of safety factors but they also have the incidental benefit of reducing overcrowding.)

Social Cost

Another method of dealing with public costs is by persuasion and exhortation. The government has, for example, encouraged research on the development of bio-degradable foaming agents so that the problem of coping with detergents at sewerage works is reduced. A third method of dealing with public costs involves financial penalties. A firm which violates the Clean Air Acts can be fined. It is conceivable that the government could introduce a special tax on firms packaging beer in cans in order to meet the ultimate costs of disposal.

All these ways of dealing with public costs require some kind of governmental activity. This gives us an important clue to the nature of public costs. Public costs are all those which for some reason or another are not taken care of by the operation of market forces. Where there is a market for a good or service the forces of demand and supply are usually brought into some kind of equilibrium through the price mechanism. The price mechanism serves as a kind of rationing device which influences what we buy and what is produced by manufacturing firms and firms providing services of all kinds.

There is no such market price for clean air or 'peace and quiet'. It is likely that a lack of clean air and 'peace and quiet' will affect property prices in an area. But a fall in property prices is unlikely to lead to a policy which would diminish pollution.

The householder pays for refuse disposal and sewerage services through the rates, but the amount he pays does not depend directly upon the amount of rubbish he produces or upon the difficulty of dealing with the effluent from his drainage system. If there were a direct relationship between rates and costs there would then be some kind of corrective mechanism which might reduce the amount of beer purchased in non-returnable containers and the quality of foaming agents used in detergents.

There is another fact about public costs which is worthy of emphasis. Public costs are not taken account of, or not taken account of in the proper way, in our national income statistics. The cost of air or noise pollution tends to increase with rising incomes but there is no satisfactory way of placing a financial value on clean air or 'peace and quiet'. Where public costs do involve financial transactions these are, paradoxically, actually counted as part of our national income. The more people spend on laundry bills, the more people are employed in picking up empty beer cans in recreation areas, the more is spent by sewage disposal authorities, the greater is the national income. The difficulty is that it is not usually practicable to separate that part of expenditure which is devoted to 'defensive' expenditure of these kinds from that part which do in fact make a positive contribution to our standard of living.

Exercise

Consider the example given in the text of a crowded lift. Suppose that an eighth person enters the lift and that the additional subjective cost to each occupant is equal to y. (a) What is the private cost of the eighth persons action? (b) What is the public cost? (c) What is the total social cost of the seventh and eighth persons action in entering the lift? (d) Would you expect y to be greater than x?

Answers

(a) $x + y$
(b) $7y$
(c) $8(x + y)$
(d) Since x and y are both subjective costs there are no *a priori* grounds for saying that one is greater than the other.

Exercise

Suggest ways in which the magnitude of the subjective cost of aircraft noise might be estimated.

Answer

One method is hinted at in the text. The subjective cost of noise can be estimated by attempting to measure the effect on property prices, e.g. by comparing the market value of otherwise similar houses in different noise situations. An exercise of this kind was carried out by the Roskill Commission on the siting of London's third airport. Another method of attempting to estimate the cost would be to calculate how much money would have to be spent on sound proofing properly. Neither of these methods have yet proved fully acceptable either on practical or logical grounds.

3 Why traffic congestion occurs

The examples given above have all been selected to help illustrate different aspects of the public cost associated with traffic congestion. Smoke and hydrocarbons, carbons and nitrogen oxides from vehicle exhausts pollute city air just as fumes from the factory chimney. There are subjective public costs of the delays imposed upon traffic and upon pedestrians trying to cross the street. There are financial costs through extra consumption of fuel and wear and tear of vehicle engines which have to be met by the vehicle users. There are subjective and purely environmental costs associated with stresses on frustrated motorists and the visual intrusion of a queue of vehicles on what might otherwise be a pleasing urban scene.

PrivateCost

Public and private costs can be distinguished in the traffic congestion situation just as in the previous examples. With some items, like the delays to pedestrian movement, there are public costs which are different in nature from the private costs. In this case the costs are borne by the people who live, work, or shop in the congested area irrespective of whether or not these people themselves contribute to traffic congestion by using a road vehicle.

Public Cost

But many of the public costs involved in traffic congestion are similar in nature to the private costs—just as in the example of the crowded lift. Vehicle users share between themselves the extra costs of fuel, wear and tear, maintenance and delays which a congested situation produces. But this does not mean that there is equality between the private and public costs involved. Each vehicle imposes a delay upon all other road vehicles, and just as the example of the crowded lift helps to illustrate, the private cost to the individual vehicle can be much less than the costs imposed upon all other vehicles.

Suppose that 500 vehicles are travelling along a heavily congested street at 15 km/h, and that the costs of this part of their journey work out on average at £1·00. Suppose next that if 501 vehicles used the road average speeds would fall slightly to 14·8 km/h, and average costs are increased to £1·01. The cost to the single extra vehicle is £1·01 for the journey just like the rest. But the cost imposed by the single extra vehicle on the other 500 is £5·00 (500 × 1p). In other words, a single extra vehicle imposes a cost on other road users equal to nearly five times the average private cost. This sum of £5·00 is what economists call the *marginal public cost* of congestion. (The word *marginal* is used in economics to denote a small change. In this case the smallest change which can occur takes the form of a single extra vehicle.)

It does not matter which of the 500 vehicles we count as the 'extra' one. The point is that the marginal public cost of the single vehicle works out at £5·00. Or in other words the cost of £5·00 would be avoided if there were one fewer vehicle in the system.

People's consumption of most kinds of scarce resource is usually limited by the operation of market forces.* Many people, for example, would

* There are, of course, many examples besides traffic congestion where market forces fail to limit consumption. One kind of example, similar to that of traffic congestion, is attributable to the lack of any market such as for clean air, peace and quiet, and a good environment (as mentioned in Section 2 on social costs). But even where there is a market, such as for scarce mineral resources, market forces do not operate to reduce consumption unless the scarcity is recognized and taken into account by purchaser and suppliers.

like to live in a detached house with a large garden. But the very fact that many people would like to consume substantial amounts of residential land pushes up the price, and as a result urban living at a density of one house in a plot of a quarter of an acre or more is a privilege enjoyed by only an affluent minority.

The price mechanism which helps ration residential space does not exist for road space. The only disincentive to the motorist in 'consuming' congested road space is the private costs he bears, and as indicated this private cost may be much less than the public costs which he imposes upon other road users.

The individual usually incurs a substantial expenditure for the cost of his car, garaging, licensing, testing and insurance. But the marginal cost of running his car in terms of petrol, tyre wear and so on are usually relatively small compared with the convenience of a means of transport which conveys him (with full protection from inclement weather) from the door of his home to the parking place at his destination. In most cases the only disincentives are the delays imposed by traffic congestion itself. The individual is acting in an economically rational way in using a congested road for as long as his private costs are less than the benefits he obtains through making the journey by car. There is no mechanism by which the individual is obliged to take account of the public cost which is involved in his use of a vehicle in congested conditions.

Exercise

What exactly are the items which go to make up the private costs of making a journey by car?

Answer

The costs of actually making a journey by car consist only of running costs (petrol, oil, wear and tear) and time costs. They do not include capital cost, depreciation, licensing or garaging because these costs are incurred irrespective of whether a particular journey is made.

Exercise

Which of these costs do you think most individuals take into account in making a journey by car?

Answer

It is likely that many drivers take into account only petrol costs. Other mileage-related costs occur only at relatively infrequent intervals and so the individual does not take them into account in making a particular journey. Since time costs are subjective it is difficult to assess the extent to which they are taken into account.

Exercise

Section 3 of the unit states that social costs are ignored or not accounted for in the proper way in national income statistics. How are the different items in the social costs of congestion treated in national income statistics?

Answer

Costs which involve financial transactions such as running costs are counted as part of the national income. Time costs are also counted as part of the national income where the driver is paid as a driver, otherwise they are subjective in nature and therefore they are not included.

4 Bendtson's two theses

This explanation of the existence of traffic congestion has implicitly assumed that the private road-vehicle is the only means of transport available and that vehicle users do not have much choice about when and where they make their journeys. In fact neither of these assumptions is realistic, and it is necessary to relax them to help explain why congestion does not get worse and worse until the system grinds to a halt. Nearly all vehicle users have some alternative means of transport available to them and they can choose not to use their vehicle in congested conditions. The private cost of making a journey in congested conditions may be small compared with the social cost, but the magnitude of the private cost may still induce some would be road users to find an alternative.

There is plenty of evidence that congestion does in fact act as a deterrent. Travelling by an alternative route is sometimes practicable. Many vehicles make circuitous journeys by side roads to avoid the delays which would be encountered in the main streets. At a larger scale vehicles travel round the town by a ring route in order to avoid the congested centre. Making journeys outside peak hours is another way of avoiding congestion. Van deliveries to shops in the major cities are commonly made early in the morning before the journey to work rush has begun. Many car travellers attempt to avoid congestion by travelling before or after the peak, and in the major cities there is a gradual build-up of traffic to the peak with a subsequent decline extending over a period of several hours which is a pattern quite unlike the short and sharp peak characteristic of the small town.

Congestion also affects the origin and destination of journeys. Possession of a motor-car not only makes it possible for people to live in villages or distant suburbs; people also choose homes at a decentralized location partly because they can use their car in less congested conditions. Shops move out from the city centre to suburban centres, to smaller towns, or even to out-of-town centres. There is even a centrifugal movement in office employment. The Automobile Association, for example, has recently moved its headquarters from Piccadilly to the town of Basingstoke which is 60 miles from central London.

> The one road system is the medium for the movement of all private cars in and through London as well as for buses, freight vehicles, motor and pedal cycles. In addition, the road system is used for many and varied pedestrian activities. The network lacks vertical segregation between conflicting vehicle flows and between pedestrians and vehicles where capacity is limited, and has few specialised facilities for buses or commercial vehicles. Despite these conditions, an equilibrium between demand and supply is reached where demand for road space is diverted to other means of transport or where movement is physically inhibited.
>
> Greater London Council (1969) *Greater London Development Plan, Report of Studies*, pp. 167–70.

Use of a different means of transport is an alternative for the car user. Travel by rail is a possibility for the journey to and from the centre of the major cities. Travelling by bus is usually a less attractive alternative since the bus is much slower than the car because it has to stop for passengers as well as being impeded by traffic congestion. Where congestion is exceptionally severe walking instead of using a car can be an attractive alternative for all or part of the journey.

The existence of this variety of alternative does have consequences of considerable practical importance. The fact that the private costs of making journeys by vehicle can act as a deterrent does provide a practical limit to the severity of congestion. City centres do not actually seize up with traffic even at peak periods, at least not very often. A kind of equilibrium is reached at very low traffic speeds.

On the basis of a study of the trends in North American and European cities this equilibrium has been formally expressed by P. H. Bendtson (a Danish Professor of Engineering) as a kind of law:

Thesis No. 1

As the number of motor cars increases, the speed of motor traffic in the central area will be reduced, during peak periods, to a minimum workable speed of 8 to 10 km/h, varying slightly as between one town and another. But even if the number of cars reaches the level of 500 to 600 cars per 1 000 inhabitants, there will generally be no actual breakdown in commuter traffic.*

The reduction of traffic speeds to as low as 8 km/h (or 5 mile/h) can be avoided by traffic-management schemes which increase the capacity of the road system. In central London, for example, traffic is channelled into an extensive system of one-way streets and this makes it possible to reduce delays at intersections by computer programmed control of traffic signals.** But the gains in traffic speed are partly offset by an increase in journey length. In effect traffic is to some extent being diverted instead of delayed.† But these traffic management schemes have kept speeds in central London at peak periods up to about 18 km/h.

The growth of private vehicle ownership and use is of course mostly at the expense of the existing public transport network. There is therefore a corollary to Bendtson's first thesis:

Thesis No. 2

The share of public transport in the total traffic to and from the central area is primarily governed by the law expressed in Thesis No. 1. The proportion of commuter traffic using private cars will always reach, and remain at, the level corresponding to the minimum workable speed referred to above, and the remainder of the traffic will use public transport. To some extent, the distribution of traffic over private and public transport will also depend on the quality of the latter (e.g. whether or not underground railways are available), but the importance of this aspect is hardly as great as is often thought.

This thesis is worth stating in different words. According to Bendtson the level of car ownership and the capacity of the road system are the crucial variables in determining the modal split between cars and other means of transport. Other movement systems like public transport systems and walking cater for a residual. If the thesis is right it follows that improvements in the public transport system or in the provision of facilities for pedestrians will not by themselves have much effect in reducing traffic congestion.

* P. H. Bendtsen (1961), *Town and Traffic in the Motor Age*, p. 159, Danish Technical Press.

** The TV programme accompanying this unit describes the West London computer controlled traffic scheme.

† See J. M. Thomson (1968), The Value of Traffic Management, *Journal of transport economics and policy*, Vol. II, No. 1, Vol. II, No. 2, and Vol. II, No. 3.

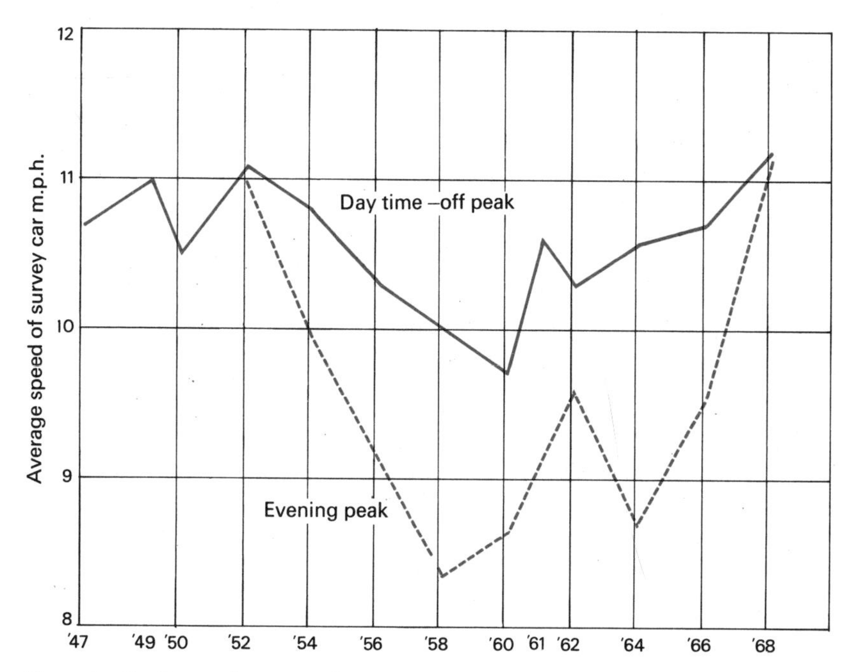

Figure 3 Traffic speeds in central London 1947–68.

Source: Greater London Council (1971) *Greater London Council Intelligence Unit Quarterly Bulletin*, No. 16.

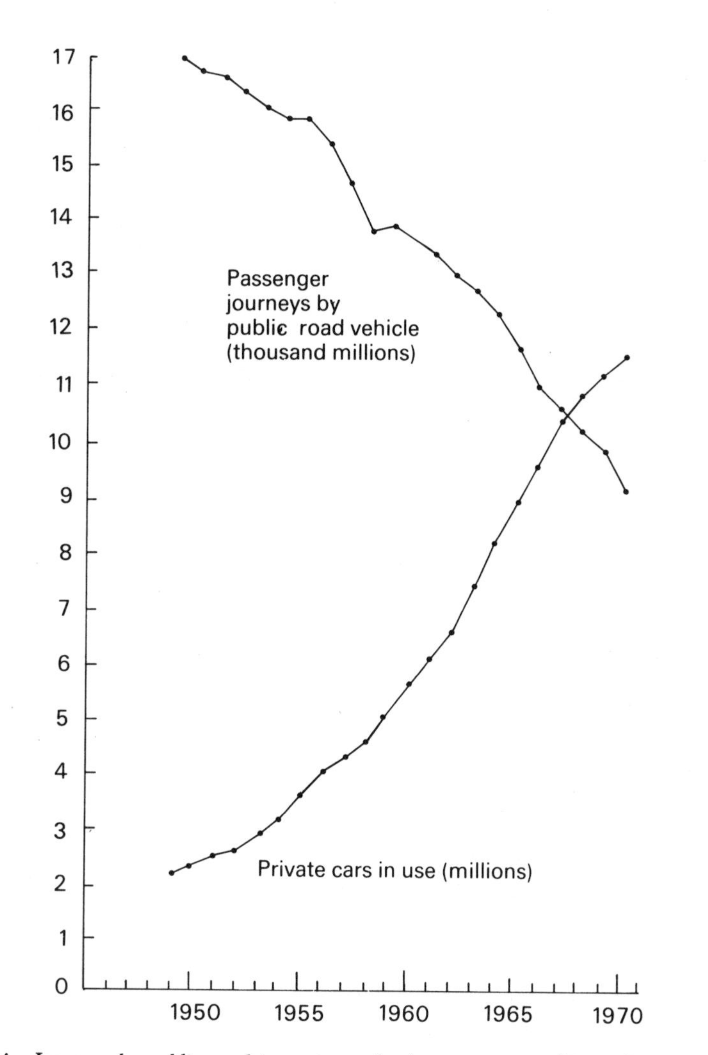

Figure 4 Journeys by public road transport and private car ownership in Great Britain 1949–71.

Sources: Annual Abstract of Statistics.

5 The speed and flow of traffic

This discussion of the economics of congestion has so far been limited to general relationships and hypothetical examples. In this and the following section the relationships are expressed in more realistic form. Consider first the relationship between the speed and flow of traffic which depends partly upon the nature of the road system.

The relationship between speed and flow is different for a motorway, to take one extreme example, and a closely knit network of streets characteristic of city centres. The relationship will be different again in an urban area where the traffic has been extensively channelled into a system of one-way streets. But in the typically congested condition of most urban roads the relationship between speed and flow approximates to:

$$v = d - fq \tag{1}$$

where v is the average speed of traffic (in km/h), q is the flow of traffic (vehicles per hour), and d and f are constants.

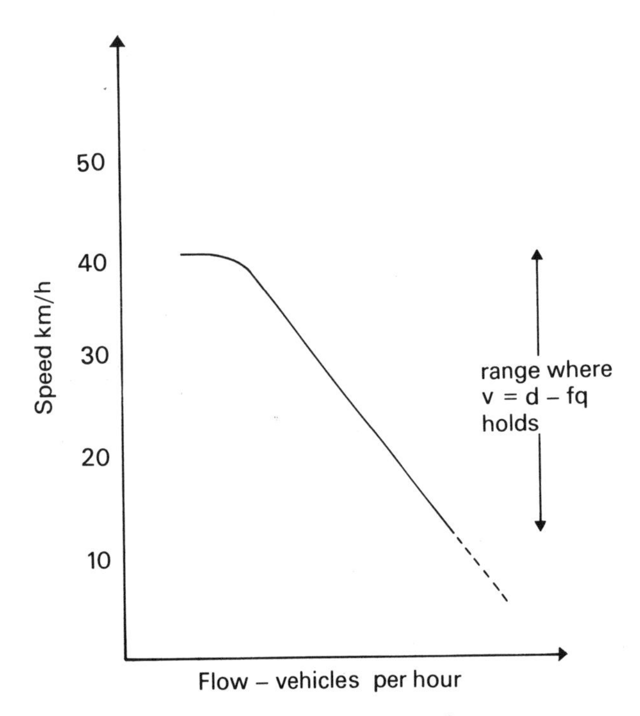

Figure 5 Speed and flow of traffic on urban road networks.

This formula does not hold for all kinds of traffic conditions. Where traffic is very light the average speed may not depend upon the volume of traffic. When traffic is exceptionally heavy there are jams and snarl ups and the relationship between speed and flow is irregular. But cases like these are not typical of a congested road network. Over the range of speeds between say 12 and 40 km/h for which data is available, this simple relationship ($v = d - fq$) appears to hold fairly consistently.

The constant d is usually interpreted as the 'desired' speed, that is the speed at which vehicles would travel if they were not impeded by other vehicles. Where q is below a certain value (depending upon the capacity of the road system), the relationship $v = d - fq$ does not hold and instead the traffic speed approximates to d. The interpretation of d as the desired speed can be used to help a give definition of traffic congestion. *Traffic congestion occurs where drivers are obliged to travel at less than their desired speed because they are impeded by other road vehicles.*

The speed of traffic depends upon the composition as well as the volume of traffic. A single bus delays traffic more than a single car does. A motorcycle impedes other vehicles less. The constant q in the equation is not simply the number of vehicles, but is the number of vehicles expressed in terms of *passenger car units* or p.c.u.s as they are often called. The p.c.u. takes into account the variation of different vehicles contribution to congestion. A bus counts as 3 p.c.u.s. A heavy goods vehicle as 2 p.c.u.s. A motorcycle as 0·75 p.c.u.s.

The constant f is a measure of the sensitivity of the road system to changes in the volume of traffic. The nature of f can be illustrated by considering what happens when there is one additional vehicle in the road system. With $q+1$ vehicles in the system the new speed of traffic is given by v_1 where:

$$v_1 = d - f(q+1) \tag{2}$$

By subtraction from equation (1) we get

$$v - v_1 = f \tag{3}$$

In other words, the speed of traffic is reduced by f when there is one extra vehicle in the system. The size of f depends in practice upon the capacity of the road system being considered. If the relationship is applied to a single narrow street f will be large since a small change in the volume of traffic will lead to a large change in speed. If on the other hand the relationship is applied to an extensive network including major roads, f will be small because a large change in the volume of traffic will cause only a relatively small change in average speeds.

Exercise

The following data relate to traffic flow and speed at peak and daytime off-peak periods for roads in the central area of Sheffield in 1967

	Peak	Daytime off-peak
Average flow p.c.u./h	1 510	1 325
Average speed km/h	11·1	16·6

[Source: M. Marlow (1971), *Repeat traffic studies in 1967 in eight towns previously surveyed in 1963/4*, Road Research Laboratory, RRL Report LR 390.]

From these data estimate the 'desired' speed.

Answer

56 km/h

Exercise

Estimate the average speed if the peak flow increased by 100 p.c.u./h.

Answer

8·1 km/h

[These results will be obtained through solving the two simultaneous equations:

$$11{\cdot}1 = d - f\,1\,510$$
$$16{\cdot}6 = d - f\,1\,325]$$

6 The marginal public costs of congestion

The private costs of making a journey by motor vehicle can be expressed by the formula:

$$c = a + \frac{b}{v}$$

where c is the cost per km and v is the speed as before.

The constant a represents these costs which are just a function of distance (like most of petrol costs). The constant b represents those costs which are a function of time spent travelling—like the cost of time itself and part of petrol and maintenance costs. Later in this section there is a discussion of exactly how specific costs should be allocated as part of a or as part of b. But for the moment this question is left open and this equation will be used to derive an expression for the marginal public cost of congestion.

The table below shows how private costs vary with speed on the assumption that $a = 1$p/km and $b = 100$p/h.

Speed (km/h)	Cost (p/km)	Change in cost for decrease in speed of 1 km/h
25	5·00	0·17
24	5·17	0·18
23	5·35	0·20
22	5·55	0·21
21	5·76	0·24
20	6·00	0·26
19	6·26	0·30
18	6·56	0·32
17	6·88	0·37
16	7·25	0·42
15	7·67	0·47
14	8·14	0·55
13	8·69	0·64
12	9·33	0·76
11	10·09	0·91
10	11·00	1·11

These values for a and b used in this table have been selected on a fairly arbitrary basis but they serve to illustrate the fact that the private costs of travelling increase sharply with decreasing speed. A reduction of speed from 23 to 22 km/h increases costs (on these assumed values) by 0·2p/km, but a reduction of speed from 10 to 9 km/h increases costs by 1·1 p/km. In fact the changes in costs for a small change in speed is equal to b/v^2. (You can check this relationship by calculating b/v^2 and writing down the result next to the figures in the third column of the table.)

An expression can now be derived for the marginal public cost of congestion, i.e. the cost of a single extra vehicle in the system:

> The change in average speed caused by a single extra vehicle or more accurately an extra p.c.u., is f (see preceding section).
> The change in costs for a small change in v is given by b/v^2.
> The number of vehicles is q.

The marginal public cost of a single extra vehicle is therefore given by:

$$m = f\frac{b}{v^2}q \qquad (5)$$

Substituting in equation (1) for q we obtain* for the marginal public cost:

$$m = f\frac{b}{v^2}\frac{d-v}{f}$$

$$= \frac{b(d-v)}{v^2} \qquad (6)$$

A striking feature of the relationship shown in this formula is that the marginal public cost of congestion is independent of the size of the road systems considered, i.e. it is independent of f. But it is not difficult to see why this is so: where we are considering a large system a lot of vehicles are affected but the reduction in speed resulting from a single extra vehicle is small: where a small system is considered fewer vehicles are affected but the reduction of speed is greater.

Figure 6 shows how the m.p.c. of congestion varies with speed on the assumption that the 'desired' speed d is 50 km/h and that b = 100 p/h. At traffic speeds of 30 km/h the marginal public cost of a single vehicle works out at 2·2 p/km. But if the traffic speed is as low as 10 km/h then the m.p.c. of a single vehicle works out at 40 p/km.

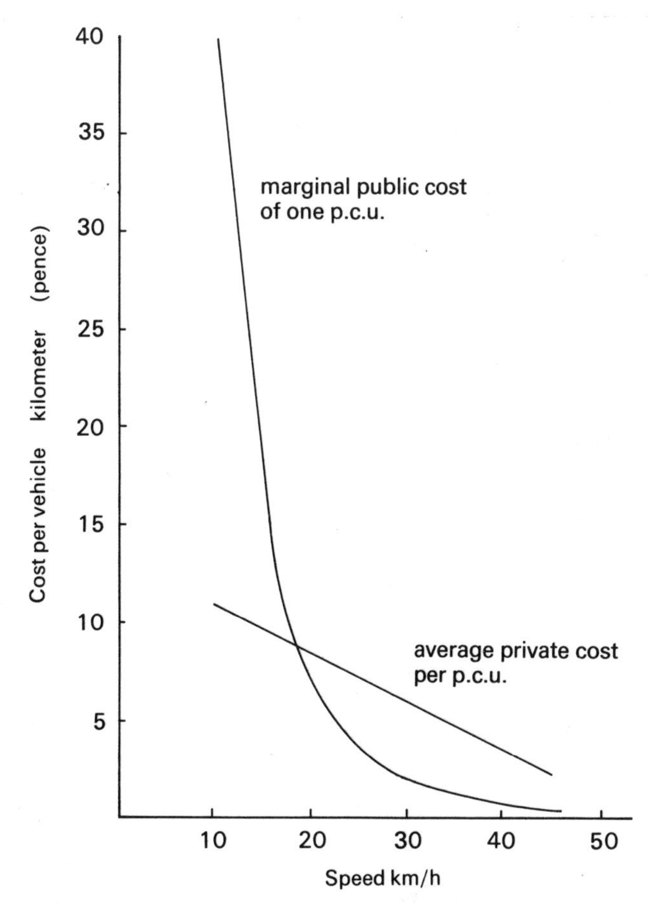

Figure 6 Marginal public cost and private cost at different traffic speeds.

* Mathematically this expression may be obtained by differential calculus:

$$m = q\frac{dc}{dq} = q\frac{dc}{dv}\frac{dv}{dq} = \left(\frac{d-v}{f}\right)\left(-\frac{b}{v^2}\right)\left(-f\right) = \frac{b\,(d-v)}{v^2}$$

For the sake of contrast the diagram also shows how private costs vary with speed on the assumption that $a = 1$ p/km. Although both private and the marginal public cost increase with decreasing speed, the increase is much sharper for m.p.c. for speeds below about 30 km/h. The reason for the difference is that while private cost is an average which is inversely related to speed, the public cost is a marginal cost which is inversely related to the square of the speed.

So far no attempt has been made to specify exactly what is included in this formula for marginal public cost nor exactly what items are counted under the constants a and b. The first point to emphasize is that this expression for m.p.c. covers only those financial costs and time costs which users of vehicles impose upon each other. The formula for m.p.c. does not include the pollution costs associated with the emission of fumes from idling engines. It does not include any allowance for the delays imposed upon pedestrians who want to cross the road. And it does not cover such factors as noise or other environmental costs which are related to the volume of traffic.

The second point is that exactly what items should be included under the constants a and b varies according to the type of vehicle, and according to the subjective attitudes of road users. In the case of the private motorist, for example, items like garaging and insurance and a large part of depreciation costs should be excluded from the formula $c = a + \frac{b}{v}$ on the grounds that these costs are independent both of mileage and of time spent travelling. The motorist has to meet the cost of the garage whether or not he ever takes his car out of it. The value of a 10-year-old car is much the same whether it has 10 000 or 100 000 miles on the clock.

In the case of buses, by contrast, garaging and depreciation costs should be included in the constant b on the grounds that the number of buses an operator requires depends mainly upon their speed at peak periods. If an

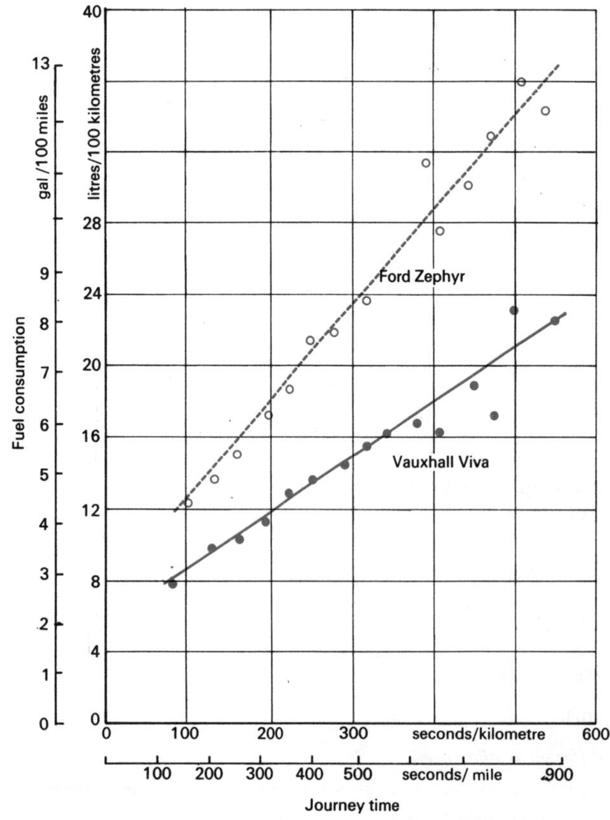

Figure 7 Fuel consumption and speeds of cars in central London.

Source: P. F. Everall (1968), *The effect of road and traffic conditions on fuel consumption*, Road Research Laboratory Report LR 226, London, HMSO.

operator has to meet peak demand on the basis of timetables which assume an average speed of 15 km/h he may need twice as many buses as he would need if average speeds were 30 km/h.

It is worth noting, however, that in the case of both buses and cars and in fact most other road vehicles a high proportion of running costs are more closely related to time spent travelling than to distance covered. An engine consumes petrol whether it is idling or whether it is providing motor power. The wear and tear of brakes, gearbox and clutch are a function of the extent to which they are used, not mileage.

The valuation of time spent travelling raises quite serious difficulties. It is usually assumed that the time of people travelling in the course of their employment should be valued at their hourly wage or salary rate, i.e. the value actually paid by the employer. But it is sometimes suggested that it would be more appropriate to value time on the basis of the value to the employer, i.e. wage or salary rate plus an allowance for overheads. This approximates to what economists call the *opportunity costs* of employing a person and is usually the basis on which an employee's time is costed in preparing a quotation for a contract.

Value of time

A study into the value motorists place on their time was carried out in Italy in co-operation with Concessioni e Construzioni Autostrade S.p.a. in 1967 and 1969. A brief report on the first part of the investigation was given in *Road Research 1968*, pp. 74–5.

The second part of the study was carried out in Milan in March 1969 on similar lines to the first part in Rome at the end of 1967. The study was based on the choice of route between an autostrade, on which a toll was charged, and a slower free road.

The basic elements of the investigation were origin-destination surveys combined with traffic counts and journey-time surveys. The surveys were carried out close to Rome and Milan and covered motorists travelling to those cities from the south over distances of up to 160 kilometres. The analysis was restricted to journeys where there was only one realistic alternative route to the autostrade and whose destination was Rome or Milan.

About 6 000 useable questionnaires were obtained, divided fairly evenly between the two halves of the investigation. The proportions of vehicles using the autostrade were related to time differences and cost differences by means of logit analysis. Journeys were divided by journey purpose—in course of work, to and from work, and other—and into the three sizes of car for which there were separate toll tariffs. Only indirect information was collected about income, but this is highly correlated with car class.

The average values of time per hour per car for different journey purposes taken under the condition of these experiments are:

In course of work	£3·00 per hour
To and from work	£0·90 per hour
Other	£1·25 per hour

These are based on an exchange rate of 1485 lire to the pound.

The average occupancy of vehicles was about 1·5 persons for work and commuting journeys and about 2·0 persons for other journeys.

The value of time varied greatly with size of car, the overall average being £1·20 for small cars, £1·55 for medium cars and £2·35 for large cars. These very high figures are about three times the hourly wage rate for the owners of small cars and about twice for the owners of medium and large cars.

Road Research Laboratory (1971) *Road research 1970* (Annual report of the Road Research Laboratory), pp. 38–9, London, HMSO.

The cost of non-working time is subjective (because its magnitude depends upon the attitude of the individual). Some people enjoy driving even in congested conditions. Others resent the time lost to other activities like work, family or hobbies. Because of this variation it is difficult to be convinced that any common yardstick is appropriate or meaningful.

A number of studies have been made in recent years, however, to establish criteria for the valuation of time for the journey to and from work. By means of surveys of the choices people actually make with regard to using different means of transport with different speeds and different out-of-pocket costs these studies provide some evidence that on average people value their own time at something like one-third of their earning rate. Such studies do not, however, provide any guidance on the valuation of time spent travelling by children, housewives and the retired, or of the variation among individuals towards the valuation of time.

There is evidence from some of these surveys to suggest that the other components of the critical constant b also depend upon attitudes.* Whether or not a motorist uses his car in congested conditions does not depend upon his actual running costs but rather upon what he perceives as his running costs. Although it is true to say that maintenance, wear and tear and petrol costs are fairly closely related to the time spent travelling it seems that many motorists make decisions about a journey on the basis of average petrol costs per mile. The motorist may be willing to endure more in the way of private congestion costs than is rational because he is unaware of the real costs involved.

Differences in individual circumstances, attitudes to time spent travelling, and differences in the way running costs are treated for different vehicles affect the magnitudes of a and b in equation (4): $c = a + \frac{b}{v}$. But the variation does not invalidate the expression we have derived for marginal public cost in equation (6): $m = \frac{b(d-v)}{v^2}$. This formula depends basically upon the assumption that the avoidable costs of congestion are proportional to the time spent travelling. Or in other words, it depends upon the kind of relationship given by equation (4), and it depends upon these private costs being averaged over all types of vehicle being delayed by congestion.

Exercise

Suppose that the 'desired' speed on a road network is estimated as 60 km/h and that the actual speed is 10 km/h. Using the same assumptions about costs as given on page 19 above, what is the marginal public cost of a single vehicle in the network?

Answer

50 p/km.

Exercise

Assume that buses carry on average thirty-nine passengers and cars one and a half occupants. Bearing in mind that buses count as 3 p.c.u.s calculate the marginal public cost of the average bus passenger and the average car traveller when the 'desired' and actual speeds are as given in the previous exercise.

Answer

Bus passengers 3·8 p/km
Car occupants 33 p/km

* D. A. Quarmby (1967), Choice of travel mode for the journey to work, *Journal of transport economics and policy*, Vol. I, No. 3.

Section 7

7 The idea of road pricing

> There is no shortage of ideas for limiting the amount of traffic in towns. As this report shows, most of them have serious shortcomings. With some of them, the cure could be worse than the disease. An early conclusion from this study was that direct road pricing seemed the most promising long-term approach to controlling the use of urban roads. We are now seeing whether a feasible system can be developed.
>
> Barbara Castle (at that time Minister of Transport) writing in the foreword of: Ministry of Transport (1967), *Better Use of Town Roads*, London, HMSO.

The wide variation in private travelling costs between different individuals and between different types of vehicle, the importance of subjective attitudes in determining the magnitude of these costs and the divergence between private and public costs at low speeds all add to the attractiveness of charging vehicle users in some way for their use of congested roads. This idea is usually called road pricing.

Suppose that the users of a congested road system were to bargain among themselves about who should use the road. Partly because of the wide variation in individual time related travel costs (i.e. in the constant b) we can assume that some users could be more easily dissuaded from using the road than others. It is conceivable therefore that those who have a strong desire to use the road at a particular time could pay something to those whose desire to travel is weaker in order to induce them to travel at a different time, to go by an alternative route, or to use a different means of transport. If the two groups can agree to a transaction of this kind then everyone would be better off than they were before. Because of the increase in speed, the costs to those continuing to use the road would be reduced by more than the amount paid to those diverted—otherwise those continuing to use the road would not have reached agreement on price. The financial benefit to those diverted must exceed their losses—otherwise those diverted would not agree to their part of the bargain.

Bargaining of this kind is impracticable, but the same effect is the basis of the idea of road pricing which would impose a tax on road users so that the private cost of their journey is increased to correspond more closely to the public cost. Such a tax would have similar effects to the bargaining situation envisaged above. The volume of traffic would be reduced, because some vehicle users would find alternatives. Speeds would increase, and the real travelling costs of all vehicles continuing to use the road system at congested times would be reduced. These vehicle users might in practice pay more. Their out-of-pocket costs would increase by the amount of the tax less the reduction in travel costs. But most of the tax paid (all except the administrative costs) would be what economists call a transfer payment. It would not involve any consumption of resources but a transfer of money from those who continued to use congested roads to the government or the local authority. The real benefit would be in the reduction of congestion.

The idea of road pricing is not so impracticable as at first sight it might seem. At its simplest it would involve requiring motorists to pay money into a meter whilst their vehicle used congested roads. At a more sophisticated level it is theoretically possible for the process to be automatic with wires under the road whose function would be to record the amount of congestion and simultaneously actuate sealed meters in vehicles using the road. Under a system of this kind motorists might be faced with a quarterly bill for their use of roads in just the same form as the one subscribers receive for their use of the telephone.

The net benefits to the community resulting from higher traffic speeds can be classed as follows:

(a) Savings in the paid working time of persons who travel in working hours, including the crews of commercial vehicles and buses.
(b) Other time savings, including time travelling to and from work, shopping, etc.
(c) Savings in fuel and other vehicle running costs.
(d) Greater productivity from buses and commercial vehicles in so far as the same number of journeys can be made by fewer vehicles, requiring less capital investment.
(e) Losses to people who as a result of the price changes refrain from making journeys which they would otherwise have made.
(f) Gains and losses to other road users such as pedestrians and cyclists.
(g) Changes in the costs caused by accidents. It is not known how a reduction in congestion would affect either the number of accidents or their severity.
(h) Changes in the costs imposed by road users on the rest of the community by way of fumes, noise, direct and other undesirable features of motor traffic.

The first five classes (a) to (e) are to some extent measurable and would yield net benefits of between £100 million and £150 million a year if a meter system of the sort described were adopted under present traffic conditions. The total, of which about 7 per cent would accrue in Central London, is comprised approximately as follows:

(a)	Savings in paid working time:	+ 40 per cent
(b)	Other time savings:	+ 63 ,, ,,
(c)	Vehicle running costs:	+ 7 ,, ,,
(d)	Capital savings:	+ 10 ,, ,,
(e)	Losses:	— 20 ,, ,,

The remaining three classes, if they were measured, would probably add significantly to the total benefit. With each year that passes, if congestion continues to grow wider and deeper, the potential benefits from improved pricing will rise at a steeper rate than the rise in vehicle-mileage.

Ministry of Transport (1964), *Road pricing: the economic and technical possibilities* (the Smeed report), London, HMSO.

A government committee, chaired by Professor Smeed (then head of the Traffic and Safety Division of the Road Research Laboratory), made a detailed study of the practicability and desirability of road pricing in 1963.* The Smeed committee investigated six different meter systems for implementing road pricing and concluded 'there is every possibility that at least one of these proposals could be developed into an efficient charging system and could yield substantial benefits on congested roads'.

Road pricing

Equipment. A laboratory-scale trial of both on- and off-vehicle equipment is in progress. Development models of a vehicle identification unit developed by the Plessey Company have been fitted to a number of laboratory vehicles, some of which are also being fitted with production prototypes of a battery-operated on vehicle meter developed by McMichael Ltd.

* Ministry of Transport (1964), *Road pricing: the economic and technical possibilities* (the Smeed report), London, HMSO.

The passage of the vehicles over an 8-lane array of aerials buried in the road surface at a laboratory exit is being recorded both on and off the vehicles to give a direct comparison of the efficiency of each of the principle road pricing methods. The off-vehicle records also permit the analysis of data-transmission errors.

The vehicle units similar to those being used in this trial are also being subjected to controlled environmental tests at some commercial test stations and by the Electrical Quality Assurance Directorate of the Ministry of Aviation Supply.

Production prototypes of the clockwork-driven-on-vehicle meters developed by Salford Electrical Instruments and Smith Meters Ltd will be available shortly for testing during the laboratory trial and at the environmental test stations. Automatic meter-reading equipment intended for use with the on-vehicle meter supplied by Smith Meters will be evaluated at the same time.

Contracts have been placed with Plessey Company to complete the engineering development of a production prototype vehicle-identification unit. Mackintosh Components Consultants Ltd have been engaged as an independent assessor of the probable costs of this unit in large-scale production at some future dates.

The testing of long-life batteries for use in the battery-operated on-vehicle meter has been in progress for about two years. All systems are being subjected to repeated yearly cycles of a regime that simulates seasonal variations in air temperature plus occasional excursions to 90°C for short periods during the summer quarter to represent the effects of direct radiation from the sun on equipment installed in the window of a vehicle. Negligibly low failure rates were recorded until the beginning of the sixth (summer) quarter. From this point serious deterioration has occurred in all primary systems being tested.

The solid electrolyte battery ih a completely new development in this field and it is attracting increasing attention with its promise of very reliable performance for up to 10 years at a reasonable cost. Samples of a solid electrolyte battery developed in the USA (Gould Ionics Inc.) are being purchased for evaluation.

Road pricing would reduce rather than eliminate traffic congestion. The costs imposed upon diverted motorists would be balanced against the benefits of reducing the travel costs of those continuing to use the road system at congested periods. The Smeed committee estimated that where existing speeds were 10 mph (16 km/h) the optimum tax would be about 0·83p per minute (2 old pence), that this tax would deter 20–25% of vehicle users and that speeds would rise to 14 mph (22 km/h). Where existing speeds were 14–15 mph (22–24 km/h) the optimum tax would be 0·42p per minute, that this tax would deter 15–20% of vehicle users, and that speeds would rise to 17 mph (27 km/h). These figures help to show that the benefits from road pricing are greatest in conditions of severe congestion. The Smeed committee concluded that there would be little purpose in implementing road pricing on roads where traffic speeds are more than 20 mph (32 km/h). The committee estimated that the net benefits of such a system of road pricing would, allowing for the value of time savings, amount to £100–150 000 000 per year.

Exercise

Consider again the hypothetical example of the crowded lift given in Section 2. The idea of road pricing can be applied to this situation by imposing an admission charge for use of the lift. State what condition the optimum admission charge must fulfil.

Answer

It was assumed that with six persons the lift is not crowded. If seven people wish to use the lift then the optimum charge is that which would deter just one person.

Effects of road pricing and parking charges on the traffic in a network

A computer-based network model (RRL TAP), which employs transportation planning methods for the study of the restraint of urban traffic, has been developed. It contains a detailed description of the road network to be considered, together with the origins and destinations of trips. Route choice is modelled on the basis of minimum generalized cost. The components of generalized cost included are time, vehicle-operating costs and tolls. The speed on a given link depends on the flow of traffic according to an assumed linear relation, the constants of which can be varied to represent various classes of link. Hence the process of finding minimum generalized cost paths must be iterated until the resulting flows are consistent with the speeds used to define generalized costs.

The number of people wishing to travel between two points is assumed to be described by a demand curve based upon the generalized cost of the journey. When the equilibrium pattern of traffic flow has been reached this cost is unique, although more than one route may be utilized.

If tolls in the form of pricing of particular links or parking charges on arrival at given points are applied, a new equilibrium pattern of traffic emerges, the effect on traffic depending on the shape of the demand curves used. The resulting economic benefits can be derived directly from these demand curves and the number of trips; the resulting revenue can also be evaluated.

By modelling a situation for which data are available, a partial calibration of the model is provided and one point on each of the demand curves is fixed. With the passage of time and the introduction of new parking charges, it is possible that a second point might be derived, but present work allows for variations in demand-curve shape by treating elasticity of demand as a parameter to be inserted at perhaps three levels.

For a large network, the model involves considerable computer running time in spite of the use of special techniques. Therefore, convergence studies and investigation of general trends are being done with a small hypothetical network of 390 links and 146 intersections with trips between all pairs of points selected from 41 of these intersections, i.e. comparable to a medium-sized town. A larger network of 400 links and 200 origin-and-destination points is being used for a limited number of runs appropriate to London.

Road Research Laboratory (1971), *Road research 1970* (Annual report of the Road Research Laboratory), pp. 34–8, London, HMSO.

Section 8

8 Subsidizing public transport

To subsidize public transport in congested areas is in principle quite consistent with the idea of road pricing. Instead of deterring vehicle users through special charges, subsidies could be used to induce motorists to use public transport instead. But as far as the usual forms of bus transport are concerned it seems unlikely that a reduction of fares would have very beneficial effects. Although buses are much more efficient than cars in their use of road space they are just as subject to road congestion as other road vehicles, and, because they also have to stop for passengers they are much slower than private cars. It seems doubtful whether they would provide an attractive alternative to many motorists even if bus fares were reduced to zero.

The major example in this country of subsidies to public transport designed to reduce traffic congestion is the construction of the new underground lines in London. The first such line to be completed is the Victoria line which runs from Brixton through central London in a north-easterly direction to Walthamstow. A government committee estimated in 1959 that this line would cost £55 000 000 to build and that at the existing level of fares would result in total annual loss to London Transport of £200 000 per year.* (The line itself may be profitable but because many passengers would be drawn from other parts of the underground and bus systems the net effect on London Transport was estimated at this £200 000 loss.)

Which view is correct?

We do not think that the purposes achieved by the Victoria Line could be met by the more intensive use of buses or by the present planned schemes for physical road improvements. But services in the central area are already badly handicapped by traffic congestion and tend to run irregularly. A more intensive bus service would still suffer from these handicaps. Also, bus services are necessarily slower than the Underground, and are more expensive for regular travellers over longer distances. They do not therefore attract the peak hour passenger travelling over longer distances—indeed, they are not primarily designed to do so, but to cater for the short distance traffic and to act as feeders to and from the railways which are ideal for long-distance travel. The most urgent improvements planned for London would, at best, affect only marginally the passengers who might travel by Victoria Line. This emphasizes the fundamental fact that road and rail serve different traffic needs.

Ministry of Transport and Civil Aviation (1959), *The Victoria line* (Report by the London Travel Committee), p. 31, London, HMSO.

. . . other transport-users save time because Victoria Line lessens congestion and speeds up traffic on the rest of the Underground (where fewer passengers means shorter station stops and so more trains can be run per hour), British Railways and the roads. We have estimated that time saved by Underground traffic diverting to the Victoria Line is worth £378,000 p.a., remaining constant over the Victoria Line's life; by British Railways traffic diverting to the Victoria Line—£205,000 p.a., increasing at 1½ per cent p.a., compound over the life of the Victoria Line; by bus traffic diverting to the Victoria Line—£575,000 p.a. constant over the life of the Victoria Line; by motorists—£153,000 at 5 per cent p.a. compound for the first 15 years and 2 per cent p.a. compound thereafter; pedestrians diverting—£20,000 p.a. at 1½ per cent

* Ministry of Transport and Civil Aviation (1959), *The Victoria line* (report by the London travel committee), London, HMSO.

compound. No estimate has been made of time savings on the rest of the Underground or to traffic remaining on British Railways (together they are unlikely to be less than £10,000 or more than £30,000). Time savings to remaining road-users, including bus traffic, have been estimated at £1,883,000.

C. D. Foster and M. E. Beesley (1963), Estimating the social benefit of constructing an underground railway in London, in, *Journal of the Royal Statistical Society*, Series A, Vol. 126, Part 1, pp. 53–4.

Construction of the line was justified partly in terms of the relief of traffic congestion it would afford; this justification was confirmed by an independent study made by two economists who, taking into account 'social' items like time savings, concluded that 'there would be some net increase in national wealth as a result of the investment'.* Anyone who regularly uses the Victoria line appreciates the contribution it makes to ease of travel in London. But it is also true that anyone using a vehicle on the roads closely following the route of the line fails to notice that congestion is any less severe than in other parts of London. It seems that the studies of the Victoria line were based upon the assumption that 'other things being equal' it would relieve traffic congestion. In fact other things have not remained unchanged. Vehicle ownership has increased substantially since 1959. The imbalance between the private and social costs of using a vehicle in the congested streets closely parallel to the line is just as great as in other parts of the road system. The net effects of building the line have not been to significantly reduce congestion but rather to increase the total capacity of London's public transport system.

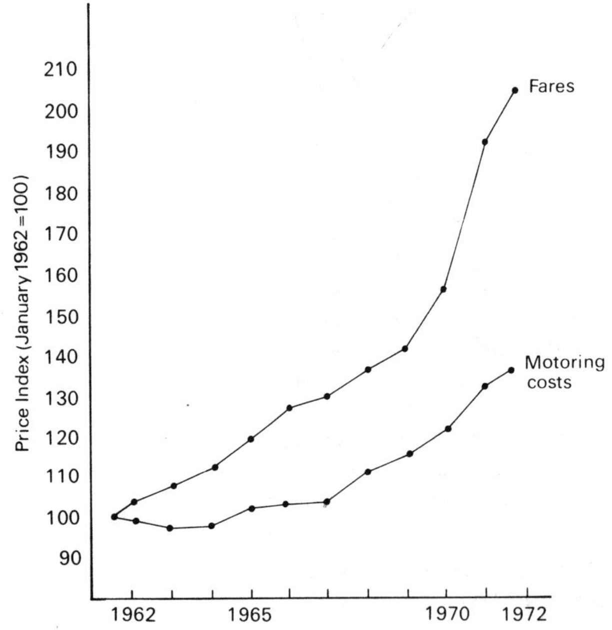

Figure 8 *Public transport forces and car running costs.*

This should not be read to mean that the Victoria line should not have been constructed. It may well be that the benefits of the increase in capacity of London's transport system outweigh the costs of the line. Whether or not this is so would be very difficult to establish. But as far as the central question of this unit is concerned the studies made to help justify construction of the line and subsequent developments illustrate the limitations of a policy of subsidizing public transport in an attempt to reduce road congestion.

* C. D. Foster and M. E. Beesley (1963), Estimating the social benefit of constructing an underground railway in London, *Journal of the Royal Statistical Society, Series A*, Vol. 126, Part I.

Section 9

9 Bus priority schemes

The success of the Victoria line in attracting passengers illustrates that public transport can compete with the private car where the road system is congested. The vital ingredient in this success is the ability to convey passengers at speeds which are comparable if not faster than door-to-door travel by road vehicle. The success depends upon the fact that the operator, i.e. London Transport, has control over the track on which the rail vehicle operates.

There are powerful arguments in favour of adopting the same kind of solution for bus transport. The average bus in central London conveys thirty-nine passengers at peak periods. The average car carries one and a half persons. If we take into account the fact that buses consume about three times as much road space as cars it can be calculated that buses are on average eight or nine times as efficient as cars in their use of road space.

Suppose that it is practicable in some way to segregate buses from other road vehicles. If buses were given enough space they would offer a reliable and speedy service and it seems likely they would attract many car owners as passengers. Other road vehicles might be more congested initially, because some road space would be occupied exclusively by buses, but if substantial numbers decided to travel by bus instead, the road space for private vehicles would be adequate and congestion would be relieved.

The theoretical advantages of such a solution are illustrated in the table below. It is assumed there are 10 000 travellers and that initially two-thirds of the travellers go by bus and one-third by car. It is then assumed that one-quarter of the total road space is reserved for buses. Next it is assumed that as a result of giving buses this extensive priority one-tenth of the car travellers would switch to the buses, and that the number of buses is increased to keep average loadings unchanged. The net effect would be that the volume of traffic on the part of the road system reserved for buses would fall by a fifth, *and* the volume of traffic on the other three-quarters of the system would also fall marginally.

Effects of segregating buses from cars

	Buses	Cars	Buses and cars together
Initial situation			
Number of persons carried	6 667	3 333	10 000
Number of vehicles	171	2 222	2 393
Passenger/car units	513	2 222	2 735
	p.c.u.s per unit of road space (initial situation = 100) on part of road system reserved for:		
Effects of bus priority scheme	**Buses**	**Cars**	**Average**
If one quarter of road space is reserved for buses	75	108	100
If (in addition) one tenth of car commuters change to buses	80	98	93

This hypothetical example is given to illustrate the idea and the effects of the relative magnitude of the different factors involved. In practice of course it is not easy to reserve a part of the road system for the exclusive use of buses. Space at the kerbside cannot easily be reserved for buses because it provides the only access for all kinds of vehicles to the adjacent buildings. It is difficult to reserve space for buses at intersections (which are the bottlenecks of congestion) without restricting the freedom of movement of other road vehicles.

But it is also true that there are many changes which can be introduced to give special priority to buses. Buses can be given the privilege of turning at intersections where turns for other vehicles are barred. Buses can be allowed the privilege of travelling against the traffic flow in a one-way street. Traffic signals can be adjusted to favour buses. Road improvements can be specifically designed to give priority to buses by means of flyovers or tunnels. All these kinds of solution have in fact been adopted in cities in Britain.

The major uncertainty connected with bus priority schemes is in their effectiveness in reducing congestion. Because most of these schemes are still on a slight scale they can only be expected to have a slight effect in attracting new passengers and thereby reducing congestion. Because the effect is small it is difficult to measure accurately, particularly in a period when car ownership is rising at a rate of more than 5% every year. Lack of knowledge is a major impediment to more extensive development of bus priority schemes; because of the uncertainty of the effects it is not easy for traffic authorities to justify large-scale restraint of other traffic in order to provide speedy and reliable bus services.

> Steps will be taken to make public transport more attractive by:
>
> (a) enabling buses to reach and circulate in the heart of the town centre as freely as possible;
> (b) giving buses priority over other traffic;
> (c) allowing bus routes through areas from which other through traffic is discouraged; and
> (d) reviewing the siting of bus stops and providing better facilities at them.
>
> Ministry of Transport circular to local authorities, Ministry of Transport (1968) *Traffic and transport plans* (Road Circular 1/68), p. 8, London, HMSO.

Exercise

Calculate the number of buses and the number of cars in the hypothetical example given in the text after one tenth of the car travellers have changed to going by bus.

Answer

180 buses (rounding up from 179·5) and 2 000 cars.

Exercise

Calculate the number of p.c.u.s per unit of road space (initial situation = 100) assuming that 20% of car travellers switch to travelling by bus.

Answer

Number of p.c.u.s per unit of road space on part of system reserved for buses is 83 (rounding up from 82·5). Number of p.c.u.s per unit of road space on part of system reserved for cars is 87. In both cases these are *indexes* taking the initial situation for the number of p.c.u.s per unit of road space as equal to 100.

If you have difficulty with this concept of *p.c.u.s per unit of road space* assume that the number of units of road space is 2 735, so that the quarter of the road space reserved for buses consists of 684 units and the three-quarters reserved for cars consists of 2 050 units. The indexes for the volume of traffic can then be calculated directly by dividing the number of p.c.u.s by the corresponding figure for units of road space and expressing the results as a percentage.

10 Car-parking controls and charges

In the 1950s it was widely believed that the major cause of traffic congestion in this country was the car parked at the kerbside. Many detailed surveys of parked vehicles were made particularly in inner London. Many new off-street car parks were established often upon the derelict bomb sites which still existed near most city centres. And, largely on the basis of successful experience in the USA, the parking-meter system was introduced to Britain.

It might be thought that the practice of charging motorists for their use of kerbside space was to some extent inspired by the idea of road pricing. But this is not so. Parking meters were introduced as a convenient means of controlling the demand for parking. The revenue resulting from a parking-meter system is partly offset by the costs of the meters themselves, but parking meters were introduced because they are easier to administer than alternative systems of control.

The principle of charging for the use of road space does, however, allow for the possibility of using the level of charges as a means for restraining demand. Parking charges can at least be used to reduce that element of congestion which is attributable to motorists cruising around looking for a place to park. A Ministry of Transport report in 1967 advocated:

> adjustment by local authorities of time limits and charges for on and off street public parking space, so as to equate demand with available capacity, and to create sufficient unoccupied space for motorists to be able to find a place to park without cruising around.*

Since the 1950s policies on parking in the central areas of major cities have also changed in other ways. It is now recognized that parking spaces can generate traffic and so add to the problem of traffic congestion, and current policy in major city centres include restraint of traffic through control over the provision of parking spaces. Current planning regulations in London, for example, now establish a maximum as well as a minimum provision of parking space in new buildings.

Policies in the 1950s

The suggestions we make will virtually solve the present problem in Inner London but, unless active steps are taken to secure that new or reconstructed buildings have sufficient parking space to cater for those who work in or visit them, a new problem will gradually develop.

HMSO (1953), *Report of the working party on car parking in the inner area of London*, No. 2, London, HMSO.

Policies in the 1960s

Provision for parking space in new developments will in general be limited to that needed for the loading, unloading, and waiting of commercial and servicing vehicles generated by the land use in order to maintain adequate control over the balance between the long and short term use of parking space.

* Ministry of Transport (1967), *Better use of town roads*, Report of a study of the means of restraint of traffic on urban roads, p. 32, London, HMSO.

Ministry of Transport circular to local authorities, Ministry of Transport (1968), *Traffic and transport plans* (Roads Circular No. 1/68), p. 8, London, HMSO.

The most promising method of restraint, at least for the shorter term, would be to intensify control over the location, amount and use of parking space, on- and off-street, especially in order to restrict long-term parking, which is a characteristic of car commuting. In some places control over the use of publicly available parking space may not be enough, and might need to be extended to privately available parking space; this might be costly and could require legislation.

Ministry of Transport (1967), *Better use of town roads*, pp. 3–4, London HMSO.

Any system of charging for parking space with the aim of reducing the severity of congestion suffers from a number of defects. First, vehicles may be moving in the area but not parking, i.e. through traffic is not affected by parking charges. Secondly, it would be difficult or impossible to apply the tax to private parking space. Thirdly, it would be difficult to justify parking charges on public space except on the basis of time rates, i.e. in terms of pence per hour. This would deter long-term parkers as much as short-term parkers, although it may be that short-term parkers contribute more to congestion. Fourthly, it may be difficult to apply parking charges to commercial vehicles.

In spite of all these defects parking charges, however unpopular they are with some motorists, are at present the only widely used measure which does something to reduce the margin between the private and public costs of using vehicles in congested areas. As such they have already achieved some success in reducing congestion. Because of this success it seems likely that they will be used more and more extensively in the future.

11 Finance and equity

Fiscal policies like charging for car parking and road pricing may be successful in reducing the number of vehicles but they can be regarded as inequitable in that private vehicle users who are deterred may not be compensated in any way. Those who switch from car to bus because of the parking charges, for example, are helping to reduce the severity of congestion, yet their new private costs (including an allowance for the extra time and inconvenience) of travelling by bus is likely to be greater than the private costs they originally incurred in travelling by car. In terms of equity therefore it would be quite logical to use the financial proceeds of a parking meter system to improve the facilities for travelling by other means. Use of the proceeds of a parking-meter system to improve the public transport system is also logical in terms of efficiency. If improving public transport induces more people to switch from car to bus then the avoidable costs of congestion are correspondingly reduced.

It is often asserted that the proceeds of all kinds of motor taxation should be used to improve the road system. The difficulty with this argument is that these taxes are not related in any way to the kind of road improvements which are needed. The motorist travelling on uncongested minor roads in rural areas may with some justification claim that his motor taxes are a subsidy to other members of the community, but it is also true that there is no great need for new roads of these kinds in rural areas. At the other extreme the high cost of road improvements in the inner areas of major cities may far outweigh the motor taxes paid by their users. And it is unlikely that road improvements in the inner areas of cities will significantly reduce the severity of congestion.

The Road Fund

The Road Fund was instituted by Mr. Lloyd George in 1909, under the Development and Road Improvement Funds Act, 1909. In his Budget speech, Mr. Lloyd George gave his reasons for setting up the Fund, stating that motorists would provide for finance, that the Exchequer would derive no advantage from the new taxation imposed on motorists, and that all the money collected was to be spent on roads. . . . Between 1915 and 1920, beginning as a war measure, part of the Road Fund money was diverted to the National Exchequer. . . . In 1926, Mr. Churchill, in his Budget Speech, repudiated the basis on which the Road Fund was insituted, and began his series of raids. In that year he appropriated £7 million from the capital of the Fund and also provided that one-third of the revenue from motor vehicle duties on cars and motor cycles should be diverted to the Exchequer. Mr. Churchill avowed that he did not wish to see motor transport grow too quickly, and wanted to protect the railways, while by raiding the Road Fund he had an entirely new source of revenue to the State.

In 1935 Mr. Neville Chamberlain noted that the Road Fund had £4,470,000 in hand and appropriated it for the general Exchequer. . . . In 1950 the Crick Committee on the Form of Government Accounts reported that no useful purpose was served by retaining the Road Fund. . . . Under the Miscellaneous Financial Provisions Act, 1955, . . . the Road Fund was abolished. . . .

British Road Federation (1971), *Basic road statistics*, London, British Road Federation.

The idea of road pricing is sometimes put forward as an alternative to existing forms of motor taxation. But to make this connection is an argument about equity not efficiency. It is logical for governmental authorities

to adopt a road-pricing solution on the grounds that it would bring private costs more into line with social costs. But the general level of motor taxation is a matter raising revenue—just like taxes on beer, spirits and wine. Governments need money for a wide variety of purposes like defence, education, health services, and social security benefits as well as road improvements, and there is no particular reason why they should be obliged to rely on income tax alone to meet the costs of these services.

The argument that it is equitable for motor taxation to be used to improve the road system is fairly weak at the national level. But at the local level it can be turned on its head. In every major town or city there are motorists who are deterred from using the road system by the existence of traffic congestion itself. As car ownership grows this number will increase. Measures adopted for the reduction of traffic, if successful, would involve a limitation of the volume of traffic, and the number of deterred motorists would be further increased. It is just as equitable for the proceeds of motor taxation to be used to improve the facilities available to these deterred motorists as it is to improve facilities for motorists who continue to use the congested road system.

The important question is not 'Who should pay for road improvements?' The answer to that question is settled by the fact that it is impracticable for any but governmental bodies to be responsible for such a complex task as the management of the road system. The crucial question is the one we started with—'what kind of solution should we adopt to reduce the severity of traffic congestion?'

There are, of course, innumerable ways in which the revenue could be used. It could be used simply as general revenue, in which case there would be a transfer of income from affected vehicle users to the rest of the community. It could be distributed between local authorities in a manner calculated to compensate residents in areas where road prices were highest. It could be used to subsidise public transport, thus compensating those forced from private to public transport. It could be used to reduce the existing motoring taxes, in particular those such as the annual licence fee which do little to restrain the use of congested areas.

Ministry of Transport (1964), *Road pricing: the economic and technical possibilities* (the Smeed report), London, HMSO.

Section 12

12 Conclusion

The fact that this unit is about traffic congestion should not be regarded as a denial of the benefits associated with rising car ownership. However expensive traffic congestion in cities becomes a person is more mobile for many kinds of journey with a motor-car than without. The advantages of this mobility contribute to the benefit of society as a whole—through enlarging the size of the labour market enjoyed by both employer and employee, as well as by enlarging the choice which the car user is able to exercise as a consumer, as a social being, and as a tourist.

Traffic congestion is nevertheless one of the costs of growth, and the faster the economy grows (as conventionally measured in terms of national income) the faster will be the growth of congestion. Some people believe that we have already reached the point where the costs of growth in this form outweigh the benefits.*

This unit has not been concerned with this wider question as it applies to rising car ownership. It has, for example, ignored the social costs of traffic accidents, and it has only mentioned the public cost of items like traffic noise, exhaust pollution and the restriction which growing traffic volumes impose upon pedestrian movement. The unit has concentrated primarily on an analysis of one aspect of the subject—the delays which road vehicles impose upon each other.

The unit has not attempted to give any clear prescription for congestion. The main reason for this is that traffic congestion with our existing technology is to an extent incurable. Whatever is done in the way of increasing the capacity of the road system, whatever is done to improve the alternatives available to vehicle users, and whatever is done to restrain car use, it seems inevitable that there will always be times and places where the number of people wishing to use a private vehicle will exceed the capacity of the road system.

The private car gives users the unique advantage of being able to convey the user more or less from A to B wherever A and B are situated. But the disadvantage is that many other vehicle users may also be using their cars on the same road system at the same time. Public transport has different advantages. It is more efficient in its use of space, and where it uses its own track (as with a rail system) the problem of congestion can be avoided entirely by scheduling the movement of vehicles. But public transport in this form does not have the ability to convey passengers from A to B. A comprehensive solution would combine the advantages of public and private transport. It would require an extension of one of the elements which is common in most of the measures for reducing congestion. These measures involve some transfer of decision making from the vehicle drivers to a control system external to the vehicle. A simple example exhibiting the feature is the traffic-light controlled intersection. Most of the measures for reducing congestion discussed in this unit also involve some transfer of decision making from the driver to an external control system.

An extension of this feature would be the design of a control system which would take over the guidance of *all* vehicles. Private-car users could specify their destination in the same way as they dial a phone number.

* See E. J. Mishan (1967), *The Costs of Economic Growth*, London, Staples Press.

They could then sit back in their seats and the guidance system would take over the driving function and convey them to their destination. External control of the vehicle would substantially increase the capacity of the road network by a reduction in vehicle headways. External control by computer would allow for an optimizing of the distribution of vehicles over the road network to minimize delays.

With a control system of this kind there would not be a very clear distinction between private and public transport. Most vehicles could be privately owned but use of them would involve parking costs in some form or another. It would be more convenient for many users to dial for a publicly owned vehicle which would convey them from A to B in the style of a taxi. Some of the publicly owned vehicles could be designed primarily for use by a single party. Others might be buses offering a more limited range of destinations but at a lower cost. Whatever type of vehicle was used the charge could be varied according to the degree of priority offered by the control system.

> We are nourishing at immense cost a monster of great potential destructiveness. And yet we love him dearly. Regarded in its collective aspect as 'the traffic problem' the motor-car is clearly a menace which can spoil our civilisation. But translated into terms of the particular vehicle that stands in our garage (or more often nowadays, is parked outside our door, or someone else's door), we regard it as one of our most treasured possessions or dearest ambitions, an immense convenience, an expander of the dimensions of life, an instrument of emancipation, a symbol of the modern age.
>
> Ministry of Transport (1963), *Traffic in towns* (The Buchanan Report, Report of the steering group), para. 55, London, HMSO.

Acknowledgements

Grateful acknowledgement is made to the following material used in this unit:

Text

British Road Federation for *Basic Road Statistics 1971*; GLC for *Greater London Development Plan, Report of Studies*; HMSO for *The Buchanan Report, The Victoria line, Report of the Working Party on Car Parking in the Inner Area of London, Smeed Committee Report on Road Pricing, Traffic and Transport Plans, Better Use of Town Roads*; Road Research Laboratory for *Road Research 1970*; Royal Statistical Society for C. D. Foster and M. E. Beesley, 'Estimating the social benefit of constructing an underground railway in London' in *Journal of the Royal Statistical Society*, series A, Vol. 126, Part 1, 1963.

Illustrations

British Road Federation for Figure 1; GLC for Figure 3; Ministry of Transport for Figure 7 and the table on page 6.

nology Foundation Course Units

1 Systems
2 The human component
3 Speech, communication and coding
4 Modelling I
5 Systems File
6 Mechanics
7, 8 Electricity and magnetism
9 Structures and microstructures
10 Statistics and reliability
11 Economics File
12 Automatic computing
13 The heart of computers
14 Computer systems
15 Analogue computing
16 Control
17 Measurement File
18 Modelling II
19 Atoms and molecules
20 Energy conversion
21 Power and society
22 Materials
23 Environment File
24 Energy in chemical reactions
25 Chemical processes
26 Maintaining the environment I
27 Maintaining the environment II
28 Production systems modelling
29 The production environment
30 Cities File
31 Urban transport
32, 33, 34 Design